THEORIES of SCIENCE

By

JOHN HODGSON

ISBN-13: 978-1519479174

ISBN-10: 1519479174

PREFACE

We all live life on this planet, as the planet ages so do we. As we learn more of what we learn we must find common significant values that affect us. When we choose our areas of study we all must find equalities within our studying to achieve common goals that have an impact on the earth. The earth is always a subject that we must learn from. Finding the answers to great questions of the unknown is of a great cause to everybody. In this book one can learn how to understand certain theories that arise when studying subjects that are enabling science to reach another level.

BEFORE LIFE there were: ELEMENTS
That EVOLVED

In the beginning of early life on earth every element existed and was present on earth. As elements shifted, twisted, and turned about the planet, eventually the elements joined together forming life on earth. Therefore elements were on earth long before life existed. Such elements were not compined as they wondered about the planet until over time they "evolved" and life became. As ecosystems grew, some small and some large, such plantations harbroughed into life forms. As this change became real to the earth many changes occurred on huge levels as well as small. In the element world the elements formed or attracted to one another by joining each other forming substances and over time life formed. At which huge forests, oceans, and mountains became. Vasts amounts of life formed in the oceans and forests containing the food they needed to survive became. These small cell forms adapted to there enviroment and evolved continually, growing larger and more complicated, developing over time with evolution.

EARTHS SPECIES

After life forms became on earth huge dinosaurs roamed the planet for millions of years they existed.

Moving forward as the planet seemed so plentiful a huge catastrophe occurred. In this event the entire earth was changed in ways that dinosaurs could no longer be plentiful. What happened was a huge astroid collided with the earth. In this asteroid, I believe, contained many new elements or "nutrients" that the earth needed but did not contain yet. As the asteroid "stirred" the earth, its elementes contained the eventual "nutrients" needed to form human existence. This still took a very long time to happen. This asteroid linked new elements that were needed for each evolutionary growth.

Comets still bring elements to the earth today. As the comets pass by our earth they release or bring the needed "nutriences" needed to substain our lives and maintain our existence.

As many and many more years to come the earth so will change and change with every season of weather. Subjecting its elements with pressure and forming even new species. And when in a long time the earth will find its own way for us humans to develop and change.

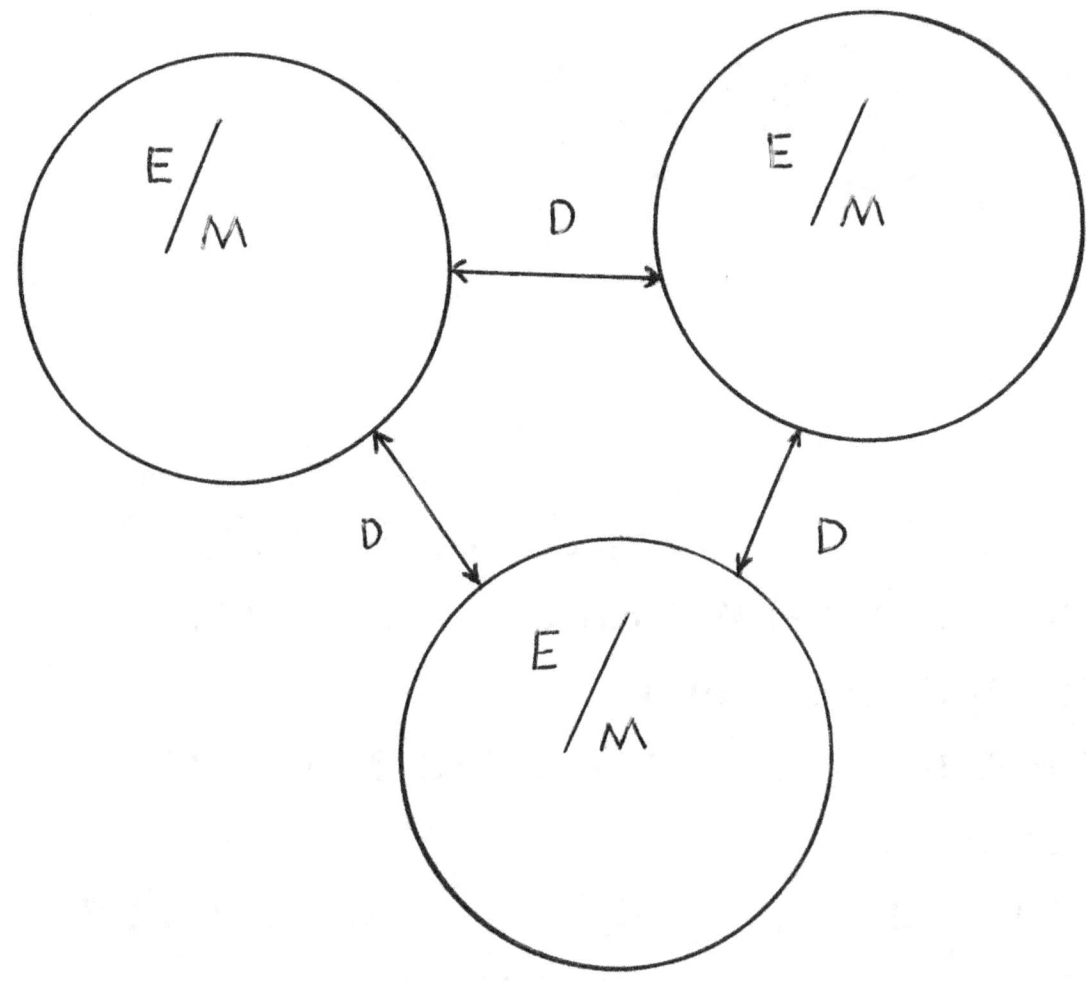

Earths and Molecules Comparison diagram.
  E=Earth
  M=molecule
  D=distance

EARTHS and MOLECULES COMPARISON

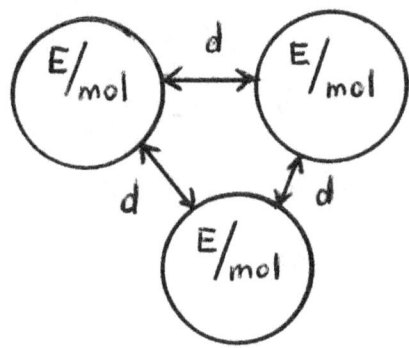

 The distance (d) between each earth (E) and molecule (mol) is a ratio as from earth to earth and molecule to molecule. The unknown is the actual measurement. If the distance between each is similar and comparable one must first understand that the distances is only a measurement and not unlike, but similar and comparable. Once the unknown material (or space) between the distances are measured then the measurement becomes a comparable between E to E ant mol to mol. Understanding that the d is explainable and the measurements are in ratio of to two

(E/mol). Then once the unknown objections are compared the equation becomes a known. Thus using each comparable as such, as a ratio, common traits are found and known matter becomes more understood. If d is a ratio of each subject of E's and mol's, then the universe is all full of other E's. And the molecules are an example of the universe right here on our earth. We can understand the universe with understanding molecules and vice versa, ultimately to understand human life, and all forms of life. To learn this is to understand human existence, cures for illnesses, and the greater cause of our lives. This is a real example of what us as humans must complete to master our survival.

EARTHS and MOLECULES COMPARISON

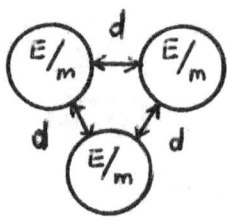

In reference from the above diagram refer to pages 3, 4 and 5.

Understanding where we could use this knowledge as a medical study one must first adjust their thinking on the equality of the subject. As to of the parameters of the theory taking place, each person is affected on a whole scale of human survival and growth. Everyone can learn this theory as outlined in on page 1. To learn how to use this theory for a growing world a person must think of the theory as a living entity, pertaining to each other and how each person can learn and provide a knowing factor of level of thought.

As the distance (d) is only a distance and comparable between molecule to molecule and earth to earth, the d is a ratio and each earth is an entity of a molecule. Molecules are understood, so by using what is understood and using a ratio of the d, being that E to E very large and mol to mol is very small; finding the common traits of principles of each the universe can be more understood. As such a mol or E may be equal to each other. Finding the common traits and learning from them the universe and molecules or earths are just a guiding factor of our existence. As the need to understand each (E/mol/universe) is of a great cause of our human growth on our earth we can lean about illnesses, and find cures of great varieties. To think this is as a pronounced inclusion that needs study is to think of this as a learning step. Each earth is of significance and each molecule is of significance in the realm of existence and study. Think if you will a molecule as an earth that is only a measurement difference, and think the distance is measured by humans. In the big picture of life sizes of these subjects is where a difference is, but other commonalities exist. So if you grasp this idea then you have stated your journey to a higher priority of learning. Remembering that there are the common traits that need to be found to captivate a higher awareness of understanding is of subject.

ATOM to ATOM and GALAXY to GALAXY

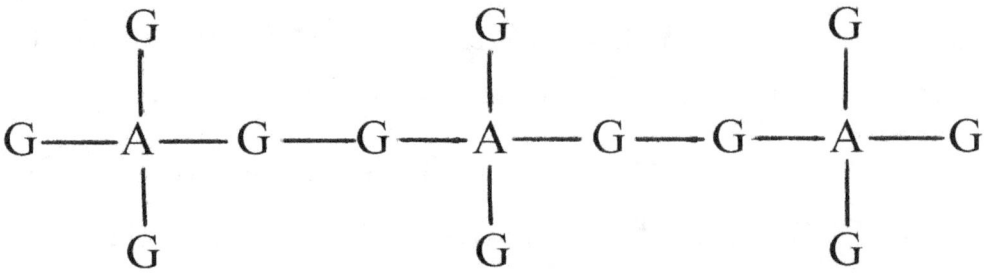

—— =bonding of a certain substance or distance between galaxies

G=Galaxy

A=Atom

Much of the bonding material is equal just as a conjoining atom to atom to form a more complex but equivalent or certain substance between galaxies to galaxies. Much of molecules or atoms bonding like galaxies bonding, I believe, are connected just on a huge scale as compared as atom to atom and galaxy to galaxy, the scale or ratio exists. Finding the correct substances holds the clue to "map" the genetics around the theory. Comparing galaxies one must ask the vast clues as compared to the existing materials and elements on earth. Each element has its own atoms and each galaxy has its own earth. Elements' atoms or atomic structure is another clue to find the code that exists in the universe as compared to galaxies between atoms (space?). The space between atoms is on a scale as galaxy between galaxies. Atoms between atoms and galaxies between galaxies are opposites in size. What's between each?

This is just a scale or ratio. The exact size is of no wonder but the most important thing is that of the clue between each. Size on the scale is the message. Expanding this theory of atom to atom bonding; if the distance between each atom is a material, then so material or "dark matter" exists between galaxies. Finding new clues exists with studying space.

These clues may be the forefront of study that can be learned to help fight disease and answers to many questions of science and outer space. Comparing the bonding of atoms as the same as galaxy to galaxy bonding. What's holding galaxies together? What's between atoms? Well the bonding material or outer space is/or may be equal to each other, only opposites in size; atoms being small; and the outer space being huge. Think if you will about the common truths between the two. Needless to say the truths exist and one must understand the truths. Once knowing the truths and equal similarities only then can we understand the magnificence that can be applided to science on our planet. Now think about the earthy factors such as molecules being joined together to form substances and such that the universe is expanding. Such molecules are expanding (or multiplying) to form many different forms of life. This is a truth that is known. With understanding the

9

galaxies significant knowledge can be understood further. Between galaxies space exists, between atoms space exists. The only known is what we can see is that which we magnify or expand. That is another clue being opposites of the same bonding theory between each atom and galaxy. To understand each of the opposites you have to find the similarities or each. The sizes are astronomacly different, but maybe the important conclusions are opposite and similar at the same existence pertaining to the common goal or the clues to find the goal traits, I believe, exist and can be found to enable us to further understand the earth and its inhabitants. Now the bonding between galaxies may be simple as gravity or pressure like the pull of the moons effects on the earth. That is a bonding understanding. Pressures exist in outer space between planets and galaxies. Furthermore earth's gravity along with the magnetic poles makes pressures on every angle of our earth. Galaxy to galaxy is maybe much similar; much as molecules bonding as an earth to earth bonds.

Find the commonalities that exist between our earth, planets and galaxies, or atoms, molecules and substances and you maybe surprised at which what you find.

FINDING SIMILARITIES that EXIST BETWEEN the UNIVERSE and LIFE on EARTH

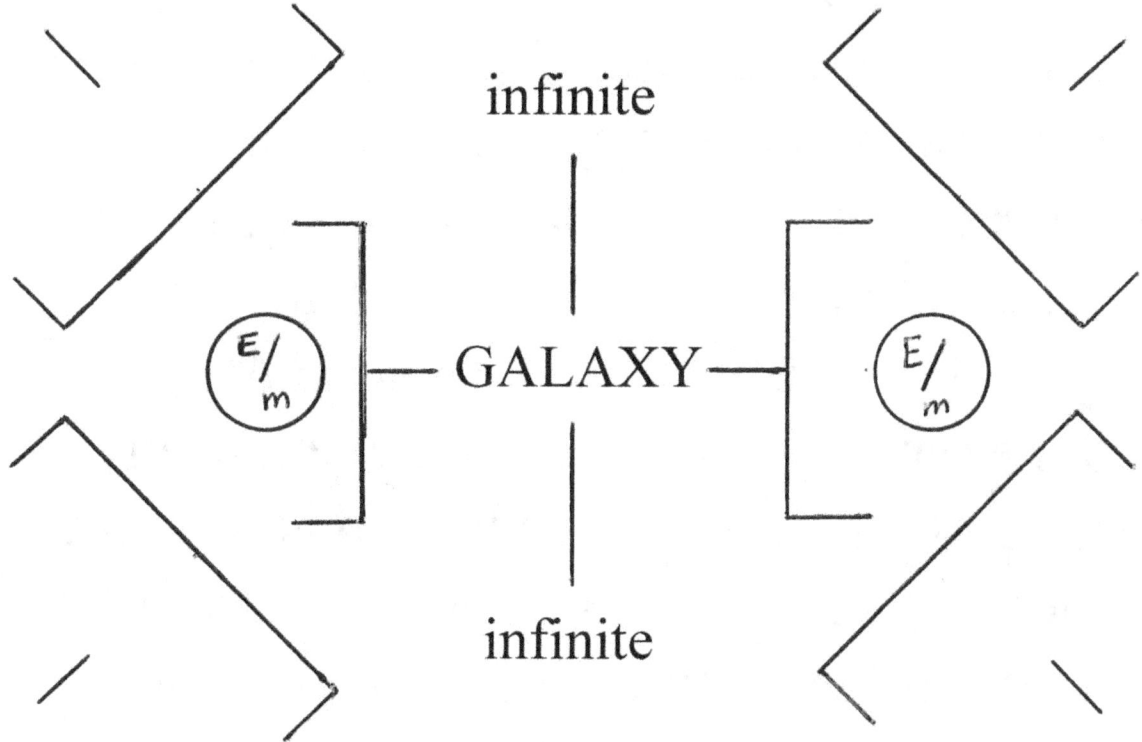

As the galaxies expand and earths are continually being created are we aware of this? They are just the same as our earth, being created and created throughout time. How long has our earth existed? For billions of years. So other earths are being created, created and created throughout time. Think of time as we understand it, sixty seconds in a minute and sixty minutes in an hour. That is how we humans perceive time. The whole universe is creating and creating on another time and space that is infinite. These creations are not yet proven but metaphorically maybe they are. Taking into account that eco-system to eco-system does exist on earth and the similarities that exist on earth could be taken into account and cannot be just coincidence, but are subject of what needs to be explored. The understanding in particular areas of space are now being found and some answers are being facts. With the understanding of the known, the unknown can be closer to understanding. The theories in question are the "similarities" not the differences that make the most sense and are facts. Finding the similarities "in" differences in the variations of the unknown are where the answers are to be found. The answers to the questions the universe holds can be applied to fundamental growth in finding clues of the unknown facts which where us on earth can be applied and eventually be more understood.

Furthermore the information available to the unknown is vast. One must decipher the knowledge and find the "congruent angles" and "parallel lines" that form in harmony.

Being that the "wealth" is great and justified as other sciences, biologically that exist, can be found and thus form new found inguinualities that are helpful for us and our planet earth.

SHADOW COMPARISON THEORY

As the suns light (warmth and radiation) warms the atmosphere it generates a rebound of radiation off the earth and reflection from the earth. The poles, north and south, are thinning and each has a hole (from gases generated by humans and the earth as its own entity.). Thus as these holes let out "steam" (or energy, heat, radiation), the entire atmosphere is affected, as this steam is released through these holes. Does this action of released steam allow the earth to cool as this 'steam' is released? Maybe. But pollution is still being made faster than it's being released. And more sun (being light and radiation) is penetrating through these holes allowing for abnormal strain on the atmosphere.

Needless to say earth's atmosphere is thinning and the moon has an impact of this. The suns radiation is growing stronger on earth. The equator is the strongest of where the warmth occurs, thus warming the entire earth. The earths crust is feeling the warmth as well. As the planet warms (by natural reasons as well as pollution) earth is its own factor of these causes. Think of the earth as an eye ball, with an outer shield and inner core. The atmosphere of an eye is the earth's atmosphere. Nevertheless as time moves forward each atmosphere wears out or ages, just like as the earth ages it changes.

As a human eye ages it wears, likewise as an earth wears. Just like congruent triangles, being different in size and time interval. Earth being at the larger size of scale or ratio, and the human eye being the smaller scale or ratio, time is of the essence. The time it takes for the eye to age is a meaning of earths aging process. This process takes time and the times are connected by a ratio. The core of the earth "erupts" and an eye can become "pink-eye". Think if you will a huge eye cascading throught the universe and that eye sees light (the sun) is the same as an earth's sun in the morning. If an eye blinks the eye has no light when it closes. When the sun sets half of the earth is dark. These are comparable. As the shadow moves across the earth it's like a blink of an eye happening. Time having a factor. It takes under one second for a blink and twelve hours for an entire shadow to cross the earth. Thus this timing is congruent to each other. As well as similar in the fact that earth is a living entity just like an eye. Such that time is a ratio. (Blink of an eye: dark rotation).

Therefore a blink of an eye is similar to the crossing shadow of the earth. When and as the crossing moves forward then the time is similar in ratio. Changes with age on earth are weather patterns and temperature. And such changes with an eye are cataracts. Other similar and congruence's can be further understood and studied. These similar occurrences may tell what might be answers to the health of an eye and the health of an earth. The cycles are time related on a scale or ratio, time being of the guiding essence. Nevertheless this is a procces governed by the bueaty of a healthy eye and a healthy earth rotating and travelling throughout the times. Large (earth) and small (eye) and with congruent attributes are pertaining to each to maintain health for both as nature so intended.

EARTH and MOLECULAR RAMIFICATION

E=M

E=earth, and M=molecule

If each M represents a molecule, then the amount existing on earth is huge. If each E represents an earth; with the expansion of the universe then the amount of earths existing is huge. On earth "new" molecules are not made, but just ever reciprocating like the three states of matter (liquid, solid, and gas) ever changing between the three states. Unlike the universe which is ever expanding, I believe, that new earths are being "formed" that are the same existence of our earth. And "new" elements (or molecules) on our planet are not reproducing only changing forms, as like clouds change to rain with temperature change. Imagine that all matter (light and dark) is not reproducing on earths, and all matter on earths are made of the same matter just changing substance (Or structure). If all matter on earth is never rejuvenating then eventually if us humans do not change our environment the earth will feel deprived and, as a living entity, need to be brought back to good heath by changing the environment accordingly.

If the universe is growing its own earths our earth is of an example of the other earths that exist to at which with study and knowledge of space our earth can and will be brought back to health, such as the earth has done in its past.

THE EFFECTS of OUTER SPACE on EARTH and EARTHS ATMOSPHERE

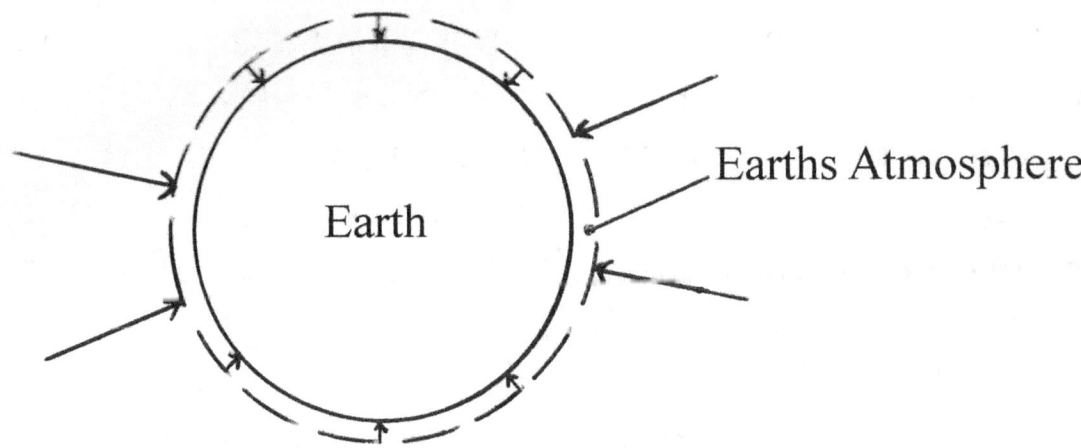

More of the green gases are concentrating on the upper ozone and are being destroying the ozone upon contact of the outer space pressures. Thus the lower ozone is generating more and more protective shield (by evaporation and photosynthesis) with green gases to accommodate the destruction of the upper ozone, as the middle ozone is staying the same to control the pressures that exist. The middle ozone must not be allowed to change too fast or the upper ozone could suffer putting pressure on the earth. Over time the entire ozone will heal. The earth must continually produce green gases, and the earth will replenish itself just as the earth has always done.

As the middle ozone is in control of the whole ozone the sun plays a huge role of the entire ozone. With the energy expelled from the sun directed upon the outer ozone the sun is reaching earth. If too much energy from the sun penetrates the earth the warmth of this energy expels and creates too much warmth. And if too much warmth is generated than the middle ozone tends to create too much ozone (or protective shield) in the middle. Then this creates too much middle ozone and the middle ozone then has to release this energy. When this energy from the middle ozone is released it moves to the upper and lower fields of the ozone. If too much of these "gases" are released the weather climates of the entire earth is effected and damaging.

COMBINATION of THEORY: HYPERLINK

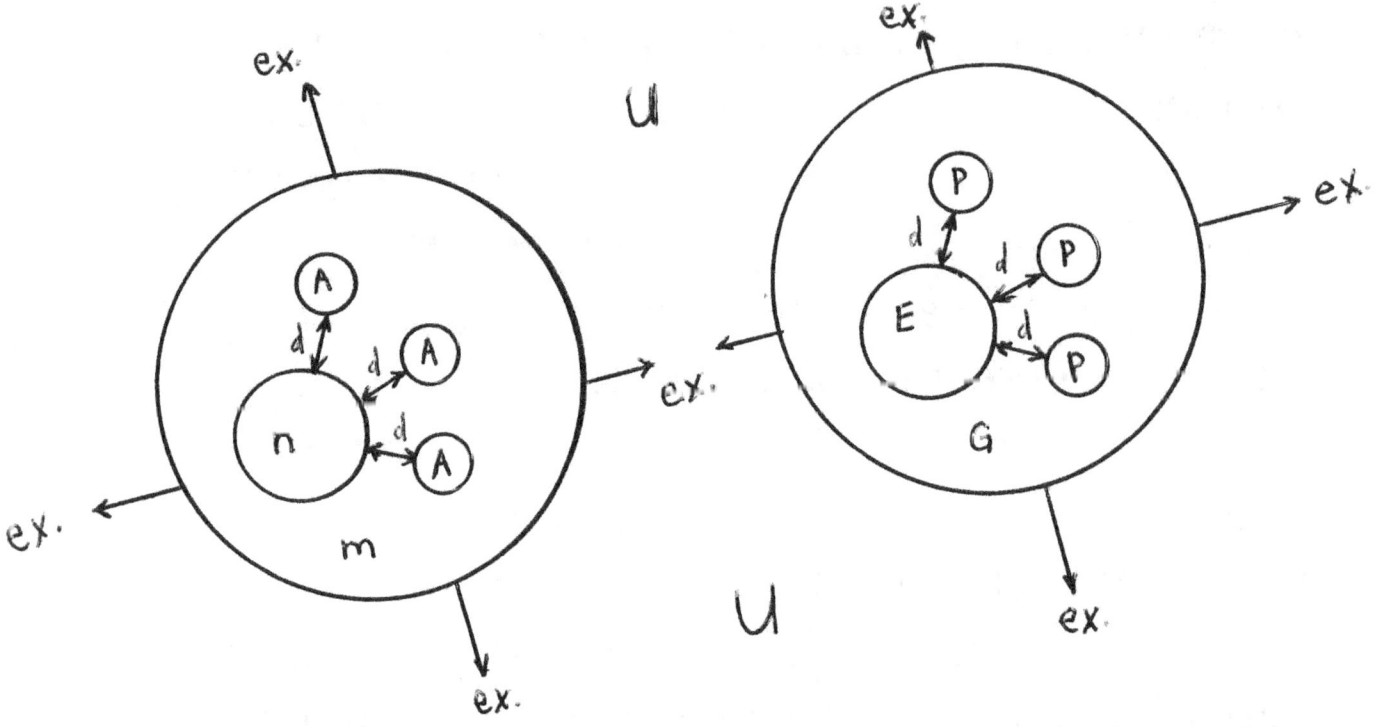

A= atom, d= distance, E= earth, m= molecule, n= nucleus, P= planet, U= universe, ex. = expanding, G= galaxy

 As the universe expands the other factors that govern the expansion are right here on earth. And can be further understood by studying the simple idea of molecule growth and the distance (d) between nucleus (n) or earth (E) and atoms (a) or planets (P).

As a molecule joins with another molecule (bonds) its like a galaxy (G) joining another galaxy and the distance (d) between the nucleus (n) and the atoms is a ratio with earths and planets. As the universe expands, galaxies are joining (or bonding) with other galaxies and with planets, just the same as atoms bonding. Think of the expansion of a growing universe at the speed of light and the distance light travels in a year. The time factor represents distance and the distance represents time. Travelling throughout galaxy to galaxy or atom to atom is just a difference in size (a ratio). Common traits exist with measurements. Thus equal bonding takes place. Nevertheless as molecules expand (or bonds with another molecules) the same is said for the galaxy governed by atoms and planets. The distance is a factor not time. And as the universe expands galaxies bonding and atoms bonding are various combinations that have similarities between each.

THE CURE for CANCER

Diagram: Timeline:

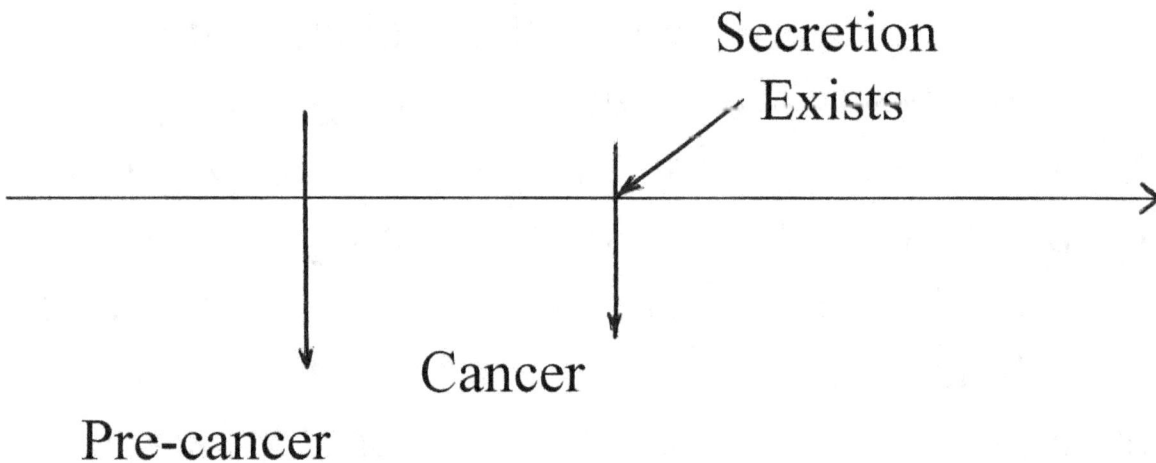

This timeline reflects the birth of cancer and pre-cancer. The liquid of pre-cancer holds the answers of the substance that is needed to cure cancer. The cure is within the disease itself.

The cure for cancer is between the cells of the body at the time of when the cancer is "born" in the non cancer cells. When the cancer is formed the "liquid" (or a secretion) of the exact time the cancer is "formed" are released into the body. These molecules are generating about the body and only at this time, when the secretion is released, the body reacts with this secretion the cure for cancer is found. These molecules or secretions are in the body's immune system. If at the time, at the very beginning of the cancer is born in the body, (refer to diagram), this molecular secretion in the body is the cure for cancer. To find this secretion one must first find the birth of the disease at which the very the cure is the secretion it's self. The secretion is only in the body for a short time after the birth of the disease. The secretion is in the cell membranes between the good cells and bad cells at the point at which they "merge", in the beginning of the cancer. To "catch" this secretion you must separate the good cells from the bad cells with experiment. Take the bad cells and place them into good cells. Using the correct forms of diagnostics methods to find out exactly when the bad cells "infect" the good cells, and then find the secretion using an elements tester, and at the exact time, this secretion can only detected for a short time in the correct cells. Then take this secretion and bottle it. Study it and apply correct other necessary elements with it and use this new liquid to apply to the bad cells and you will find the the new liquid will provide the good cells to surround the bad cells and conquer the cancer, eliminating the bad cells through their membranes. At this point the cells will, with the help of current study, cure the cancer.

Thus the cancer is a living entity and should be thought as such. The cancer has an appetite and an exhaust. Meaning the cancer feeds from the good cells and has a release such as the secretion.

Thus with modern diagnostics using experiments this secretion must be found and used correctly to be effective. The cure for cancer exists in this "time of conception" of the cancer. As the cure is found in the beginning of birth, the cure is leering just as the birth is formed.

Therefore this cancer has a food source, a digestive factor, and an emission factor, just as any living entity. The cure for cancer is in what the disease emits itself.